THAMES BRIDGES

DAVID C. RAMZAN

AMBERLEY

*This book is dedicated to my sons Alex and Toby, and my daughter Elizabeth,
who have accompanied me on many trips along and on the River Thames,
crossing the waterway on numerous occasions, using many of the bridges
recorded in this publication, especially the Queen Elizabeth Bridge, the last
crossing on the river, linking Kent to Essex where they all now live.*

First published 2022

Amberley Publishing, The Hill, Stroud
Gloucestershire GL5 4EP

www.amberley-books.com

Copyright © David C. Ramzan, 2022

The right of David C. Ramzan to be identified as the Author of this work has been asserted in
accordance with the Copyrights, Designs and Patents Act 1988.

British Library Cataloguing in Publication Data.
A catalogue record for this book is available from the British Library.

ISBN 978 1 3981 0960 5 (print)
ISBN 978 1 3981 0961 2 (ebook)

Typesetting by SJmagic DESIGN SERVICES, India.
Printed in Great Britain.

Contents

Introduction

What better place than this then could we find; by this sweet stream that knows not of the sea, this little stream whose hamlets scarce have names; this far-off lonely mother of the Thames ...

William Morris (1834–96)

From a low road bridge on the Fosse Way at Trewsbury Mead in Gloucestershire, to the towering cable-stayed Queen Elizabeth II Bridge linking the counties of Kent and Essex, the width of the River Thames is spanned by over 200 crossings on its course, including road bridges and footbridges, as well as locks, weirs and tunnels.

Many of the bridges on the upper stretch of the waterway, which is no more than a stream at this point, are simple wooden or stonework structures, then flowing to the east towards an area known as the Thames Valley, fed by many tributaries on its course, the ever-widening Thames is spanned by magnificent architectural structures erected during Great Britain's technologically inspired period known as the Industrial Revolution.

Stretching across southern England, starting as a trickle in a Gloucestershire meadow, the Thames passes through, as well as between, ten counties, inner London included, designated a county between 1889 and 1965, where on its course the river transforms into an estuary, its tidal water flowing towards the mouth of the Thames to eventually mix with the water of the North Sea.

The Thames first flowed its way eastwards from between 170 million and 140 million years ago; however, up to around 500,000 BC, from its source, the river once ran further north than it does today. When advancing glaciers of the Ice Age blocked the old river's drainage system, the Thames was gradually forced southwards, winding its way down to its present setting, where its course would define boundaries between local kingdoms and petty kingdoms, tribal lands and later the counties situated on the river.

Throughout all this time, although London Bridge has been acknowledged as the first permanent structure to transvers the Thames in the country's capital, many other bridges crossed the Thames further upstream. These were simple structures, erected for moving livestock and carts from one field to another, where the waterway dividing farmland and pasture was too wide or too strong to ford.

Ferries also provided a means to cross the Thames until innovative building techniques made it possible for carpenters and stonemasons to construct permanent structures to link communities, hamlets, villages and towns, the passage across usually permitted after payment of a fee known as a toll.

Of the 200 bridges that cross the width of this ancient river, many have a fascinating tale to tell, from their construction to events that surround them, histories dating from the first century up to, and into, the twenty-first century, where new bridges are planned to be built, linking with a modern transport infrastructure to ease traffic congestion around the capital and throughout adjoining counties.

Thames Bridges explores the economic and social history of many bridges that cross this mighty river, some unassuming, but no less important, while others are unique in their design and build.

Golden Jubilee Pedestrian Bridge and Hungerford Railway Bridge – state-of-the-art crossings of their time.

1

Trewsbury to Godstow

River, who with thy two soul-stirring names; speak'st, one of Rhedicyn's youthful dream; and one of Commerce', Empire's mighty stream; at proud Augusta's foot,-Isis, and Thames …

John Bruce Norton (1815–83)

For many years, scholars and academics have disputed where the course of the Thames begins, whether at Seven Springs, close to the town of Cheltenham, or at the recognised source, Trewsbury Mead. Although Seven Springs is further north-west and at a higher plane, in 1546, John Leland, scholar and graduate of the universities of Cambridge and Oxford, and an antiquarian and historian, wrote, 'Isis (the Thames) riseth at three myles from Cirencester, not far from a village

Trewsbury Stone.

cawlled Kemble, within half a mile of the fosseway, where the very head of the isis is.' Ever since this time most authorities have accepted Trewsbury Mead as the Thames source, relegating Seven Springs to a tributary.

At Trewsbury Mead a large stone block, erected in the 1970s, bears the inscription: 'The Conservators of the River Thames 1857–1974; This stone was placed here to mark the source of the River Thames.' The stone is situated adjacent to an ash tree, on a piece of high ground above a group of rocks, believed to be the remains of a well from where the spring flowed. This also marks the beginning of the Thames Path, a National Trail that runs adjacent to the river as it meanders through the rural counties of Gloucestershire, Wiltshire, Oxfordshire, Berkshire, Buckinghamshire, Middlesex, Surrey into the centre of London, and then beyond to the Thames Barrier at Woolwich – a distance of 184 miles. Beginning at Trewsbury Mead, a walker covering an average of 15 miles a day would take around two weeks to complete the trail, crossing many of the road bridges, footbridges and locks on the Thames, on a gradual descent from 108.5 metres (356 feet) above sea level.

For most months during the year you will find the spring at Trewsbury Mead completely dry, with water only forming during the wettest of weather when groundwater seeps upwards and the field becomes waterlogged. The water then gradually flows down the sloping meadow towards, and below, the old Roman

Thames Head Meadow.

road known as the Fosse Way, now the A433, forming into a large pool, Thames Head, in the adjacent meadow to the south-east. In the early 1960s, the road was realigned to bypass Thames Head Bridge, which once crossed the Thames and Severn Navigation, part of a waterway connecting the Severn estuary with the headwaters of the Thames.

From Thames Head, the river (known at this point as the Isis, an alternative name for the Thames derived from the Latin name Tamesis) begins running south-eastwards in the form of a narrow, shallow brook, its volume gradually increased by tributaries, some no more than trickles of water, joining the waterway as it flows downstream and under the A429, now recognised as the first bridge to cross the Thames, close to Little Hayes Farm. Passing to the east of Kemble village the Thames runs on towards the first settlement on the Thames, the little village of Ewen.

From the A429 double-arched road bridge, erected at the point where the Cirencester branch line once crossed the highway before closure as part of the Beeching Cuts during the mid-1960s, the shallow waterway passes below a minor roadway at Parker's Bridge, a low-built stone crossing with a wooden walkway linking the Thames Patch north and south of the river.

A429 bridge.

Parker's Bridge.

As the Thames turns eastwards, the river is crossed by a newly built stone bridge within Mill Farm, a seventeenth-century building in a residential revival scheme set in 5 acres on the edge of Ewen. The small hamlet was dominated by the eighteenth-century Ewen Manor, said to have been taken down from its site south of the river and then rebuilt to the north around 1730, the estate once owned by the Gibbs family, descendants of William the Conqueror.

As the Thames meanders through to the south of the village centre, it is crossed by a low bridge on the road leading out of Ewen.

Ewen Bridge.

The clear water running below Ewen Bridge is usually shallow, especially during summer months, where the highest level ever recorded reached just under a metre. From here the Thames runs on its natural course southward through farmland and meadows towards the Cotswold Water Park, straddling the Gloucester–Wiltshire border. Here a patchwork of some 133 various-sized lakes, Britain's largest water park, were created during the latter half of the 1900s by the extraction of Jurassic limestone gravel eroded from the Cotswold Hills, the pits, when all worked out, then filled with water from natural springs and rivers.

During winter months these lakes are home to over 40,000 wintering waterfowl of some forty species. At the north of the lakes is Neigh Bridge, a simple stone double-arched construction erected over the Thames west of Neigh Bridge Lake and Country Park.

Neigh Bridge.

Spine Bridge.

To the south of the river, close to Neigh Bridge, evidence of a probable Iron Age Romano-British settlement was discovered through a dense complex of cropmarks covering around 7 acres, multiple linear ditches, and indications of D-shaped and sub-circular enclosures.

Two pieces of late first- to early second-century Roman sculpture – an eagle and a shield, usually associated with a shrine – were found at the site close to Neigh Bridge.[1] Several other settlements were also located around the area of Somerford Keynes, to the north of the river, including a Romano-British settlement and a hill fort.

As the Thames transverses the lakes, flowing in a south-eastward direction below the Spine Road West bridge, several other Romano-British settlements, some of which cross the border into Wiltshire, as well as artifacts from the same period, were also discovered at the gravel workings.

1. Henig, Martin, *Roman Sculpture from Somerford Keynes Neigh Bridge* (Academia Education: 2000).

Above and below: Private access road bridges, Lower Mill Lane, south of Somerford Keynes.

The B4696.

Many of these sites were then lost during gravel excavation and the creation of the lakes and building of residential developments running adjacent to the Thames, where the river is crossed by several contemporary private access road bridges off Lower Mill Road.

Winding its way through and out of the lakes, the river is crossed by the B4696 road bridge before reaching Ashton Keynes, a village to the east of another Romano-British settlement spanning the Gloucestershire–Wiltshire border.

The land, Essitone, later known as Ashton, was held by Tewkesbury Abbey, the manor then coming into the hands of the de Cahaignes, the Keynes, the family name incorporated into the village name. Here the Thames forms many channels through the village, one branch emptying into a pool, while the main source runs along the front of several properties accessed by bridges crossing the river at Church Lane, Church Walk, Gosditch and The Derry, before running below a bridge at Ashton Keynes High Road.

The Thames turns east, below the main road leading out of Ashton Keynes then southwards towards the Swillbrook, a tributary joining the river before reaching Waterhay Bridge, where traditionally navigation on the river began. Although today the depth of the Thames at this point is very shallow, up until the late

Above: The Thames running from the north through Ashton Keynes.

Below: Access road bridges, Ashton Keynes.

Road bridge close to the Swill Brook.

1800s it was still possible for small boats and barges to navigate upstream to the bridge, which at one time was thought to have acted as a barge wharf for the community at Ashton Keynes, as well as for transporting gravel excavated from the surrounding pits since at least the mid-1600s.

The present-day plainly built Waterhay Bridge replaced a succession of masonry and stone-built bridges spanning the river, one of which, dating to the late 1800s, was made up of four attractive matching stone arches. To the north of the bridge is Upper Waterhay Nature Reserve, a large wildflower meadow laying on the Thames floodplain where the unique, creamy white or dark purple snakes-head fritillary grows. The flower, once common along the Thames Valley, now only survives in a few protected places.

Waterhay Bridge.

From Waterhay to the small town of Cricklade, founded by the Anglo-Saxons in the ninth century, the Thames is crossed by various Thames Path footbridges, farm bridges and bridges leading to private residencies, as well as the former Midland and South Western Junction Railway Bridge, part of the Thames Path, now only used by cyclists and pedestrians.

Archaeological evidence suggests this area along the Thames had seen occupation from at least the Mesolithic period, then permanently settled by the Anglo-Saxons when building a burh at the site. This burh was one of a system of fortresses and fortified towns erected in Wessex to defend against Viking raids, Cricklade being close to the Ermin Way, the main route from Cirencester.

Cricklade holds the distinction of being known as the first town on the Thames, the bridge formerly marking the ultimate limit of navigation on the river. The first bridge built to cross the Thames at Cricklade, close to the Ermine Way, was thought to have been a defensive Saxon structure.

Private bridge over the Thames, west of Cricklade.

Cricklade Bridge.

    The current bridge, erected in 1854, was built in limestone, consisting of a single arch over the river with a smaller arch to the north crossing a mill water course.

    Following the river, the Thames Path runs through Cricklade downstream on the south bank, passing a sewage treatment works road bridge, which also gives access to Cricklade Cricket Club, and on to the largest bridge to cross the Thames so far, a dual carriageway north of where the River Key joins the Thames.

A419 Ermin Way.

Eysey Bridge.

Following the general route of the Ermin Way, the A419 dual carriageway is the primary route from the M4 near Swindon towards the M5 at Whitminster. Opened in 1988, the carriageway has seen several alterations to its route over the years, and crosses the river to the east of Cricklade, the Thames flowing over a man-made concrete bed.

Taking the Thames Path on the south bank of the river, below the A419, a short walk brings you to Eysey Footbridge, built for walkers to continue their journey along the north bank, heading downstream towards Water Eaton House Bridge, a footbridge built close to a property erected on the site of an old manor house.

The road between Cricklade and Kempsford crosses the Thames at Castle Eaton Bridge, to the west of Castle Eaton village, a small settlement made up of typical Cotswold houses, many dating from the mid-1600s to the mid-1800s. The iron girder bridge with brick piers, erected in 1893, was built to replace an earlier timber bridge with stone piers and a stone causeway, which replaced an even earlier late seventeenth-century crossing. The Red Lion, situated on the riverbank adjacent to the bridge, claims to be the first public house on the Thames.

Castle Eaton Bridge.

From here, the river forms the county border between Wiltshire to the south and Gloucestershire to the north, up until reaching Redpool. In between the boundary line the Thames is crossed by Hannington Bridge on the road between Kempsford in Gloucestershire and Hannington Wick in Wiltshire. Built of stone in 1841, the bridge is made up of three oblique arches with flood arches at each causeway, which replaced an earlier wooden bridge erected during the seventeenth century.

At Hannington, which is situated many miles from the coast, it is unusual to discover an inn named after a seafarer. The Jolly Tar, formally a farmhouse, acquired the name in the mid-1800s. It was named after Captain Willes Johnson, who married into the Freke family, owners of the farmhouse, as well as Hannington Hall and the surrounding estate.

Hannington
Bridge.

Roundhouse Footbridge.

Further to the east at Inglesham stands an historic canal roundhouse, a short distance from the Round House footbridge on the Thames Path. The roundhouse, dating to the late 1700s, was one of five unusual buildings along the Thames and Severn Canal, used as accommodation for canal workers. British Waterways planned to renovate the roundhouse, situated at the eastern end of the canal, joining the Thames just above Lechlade, as part of a project to restore the canal network near Swindon.

The modern, wood-built footbridge marks the limit of navigation for a majority of motorised rivercraft; while at Lechlade, Halfpenny Bridge denotes the beginning of the navigable Thames for powered craft.

Built in 1792 to a design by James Hollingworth, the name of the bridge refers to the amount pedestrians were once charged to cross the river. The bow-backed bridge and toll house are Grade II listed, the crossing carrying the A361 into Lechlade, a manor gifted by William I to one of his Norman administrators, Henry de Ferrers.

Acquiring a market charter in 1210, the settlement of Lechlade evolved as a trading centre linked by the Thames, canal, road, and later the railway, from where over 10,000 tons of goods were handled during the early 1900s. The station,

Halfpenny Bridge.

however, closed in 1962 along with the East Gloucestershire Railway. On the south causeway the towpath passes through a separate arch, used when barges were towed carrying goods along the waterway before railways took over their trade.

In the early thirteenth century, the founders of Lechlade Priory commissioned the building of St John's Bridge, on a route between the priory north of the Thames, and Buscot to the south, from which a toll was charged before vessels could pass through.

St John's Bridge.

A charter granted in 1234 gave permission for a five-day fair to take place during August on the Gloucestershire side of the bridge, which became known as St John's Bridge.

During the late seventeenth to early eighteenth century, bargemen began complaining about the high tolls charged and refused to pay, which on one occasion resulted in the chaining over of the arch through which the barges passed. The bridge was later bypassed by way of a deep cut and lock, the highest on the Thames and one of forty-five locks on the none-tidal stretch of river.

By the early 1800s the bridge was in very poor state of repair, and after its care and maintenance was taken over by the county, a new bridge was built in 1886, which crosses the river, the lock, river channels and an island situated in between. Next to the lock house is the statue of Old Father Thames, commissioned in 1854 for the Crystal Palace exhibition. The statue was later relocated to the source of the Thames, before being moved again to its current location in 1974.

About a quarter mile downstream from St John's Bridge, the Thames now flowing into the county of Oxfordshire, a wide bend in the river, Bloomers Hole, is crossed by an arched footbridge erected in 2000. Built in steel and wood, the crossing was commissioned by the Countryside Agency on the route of the

Old Father Thames.

Bloomers Hole Footbridge.

Thames Path. On completion the pre-assembled footbridge was then airlifted into place by an RAF Chinook helicopter from Brize Norton.

The name Bloomers Hole is said to have associations to a farmer named Bloomer, who drowned at the bend in the river while attempting to ford the Thames with horse and wagon.

On the reach above Grafton Lock, the river is crossed by the wooden-built Eaton Footbridge, erected in 1936, at the site of the last flash lock on the river. Flash locks had been used throughout European inland waters since the Roman period, usually built into dams or weirs, consisting of a row of gated upright boards to maintain the river level. When the boards were removed the water was allowed to flash through, carrying the boat on the flow.

Boats passing upstream in the opposite direction were then winched or towed through the flash lock, a long, drawn out and difficult procedure to accomplish. Further west was another flash lock at Radcot, where the Thames is crossed by three linked stone-built bridges: Pidnell to the south, Radcot to the north with the canal bridge in between.

When a new cut for the Thames and Severn Canal was constructed towards the end of the late eighteenth century, which included the building of the canal bridge,

Radcot Bridge.

Radcot Bridge (reputed to be the oldest surviving bridge on the Thames, erected around 1200) now only crosses a backwater of the river.

Situated on the A4095, once an important trade route between the kingdoms of Wessex and Mercia, Radcot Bridge was on the access route to the stone quarries close to the village of Burford – the stone was also used to build Radcot Bridge. Pidnell Bridge also crosses over another channel of the Thames, where the bridge toll gate was located near the Swan Inn.

To the east of the bridge was the site of a castle held by supporters of Queen Matilda during the civil war between the reigning queen and her cousin Stephen, claimant to the English throne.

In December 1387, at the Battle of Radcot Bridge between the armies of Richard II and Henry Bolingbroke, much of the bridge was broken down and, although rebuilt in the late fourteenth century, the bridge was again severely damaged during the Wars of the Roses. The bridge was then rebuilt with a lower-profile central arch.

A short distance downriver from Radcot, another wooden footbridge – Old Man's Bridge, erected in the late 1800s – crosses the Thames at the site of a former weir and footbridge.

The next road bridge on the Thames is the Grade II listed, stone-built Tadpole Bridge dating to the late eighteenth century, carrying the road from Bampton to Burford, the settlement associated with a battle in AD 752, between the West Saxons and Mercians, the latter defeated and put to flight.

Known as Battle Edge, the confrontation was celebrated by the Burford population well into the seventeenth century, where a great dragon was constructed, the mythical beast associated with Uther Pendragon, which was then paraded all around the village. In the early 1800s, men working on a road close to the battle site unearthed a large stone sarcophagus containing a well-preserved

Tadpole Bridge.

skeleton and, although at first it was believed to have been a Saxon warrior, the remains were later dated to be from the Roman period. The coffin was then placed in Burford Churchyard, near to the west gate, where it still stands today.

Taking its name from a weir with a 10-foot lock, Tenfoot Bridge is another wooden footbridge; it connects the village of Buckland on the north bank to the hamlet of Chimney to the south. The first bridge was erected in 1869 and was replaced in 1890, and then again in 1993 to an identical design. Here the Thames Path passes by the bridge on the north side, then crosses Shifford Cut Footbridge at Shifford Lock and navigation cut, both constructed in 1898. The cut, built as a side channel of the river, then rejoins the Thames on its original course heading towards the thirteenth-century Newbridge.

Erected from stone quarried at Taynton in Oxfordshire, the bridge, one of the two oldest on the river, was built by order of King John during his reign between 1199 and 1216. The bridge became the scene of a skirmish on 27 May 1644, when Parliamentarian William Waller, making his way towards Oxford in an attempt to capture Charles I, was repelled on the riverbank by Royalist Dragoons.

Newbridge.

At one time the bridge was much longer than it is today, with many more arches, and the north span is now the only section of the bridge to cross the river. There is an inn at each end of the bridge: the Rose Revived to the north and the Mulberry Bush to the south.

The single-span Hart's Weir Footbridge, another arch-shaped bridge also known as the Rainbow Bridge, is situated close to the site of Hart's Weir, from which the bridge takes its official name.

The concrete-built bridge carries a footpath across the river from the Thames Path on the north bank, eastwards towards the Appleton Abbey. In AD 871, the abbey was sacked and although taken by marauding Danes, was later recovered by Alfred the Great.

In December 2009, Swinford Toll Bridge, which crosses the Thames between Eynsham and Swinford, just above Eynsham Lock, was sold at auction to private owners for £1.08 million after a local campaign urging Oxfordshire Council to buy it failed.

Completed in 1777, constructed by Oxford mason and builder John Townsend using local sandstone, the bridge replaced the local ferry service that belonged to the Abbot of Abingdon, Eynsham Abbey paying the ferryman for his services with bread and beer. Although pedestrians, cyclists and motorcyclists pass across for free, other motor vehicles are required to pay a toll at the toll keeper's cottage.

Swinford Bridge.

2

Oxford to Bray

There is no river in the world to be compared for majesty and witchery of association, to the Thames; it impresses even the unreading and unimaginative watcher with a solemnity which he cannot account for, as it rolls under his feet and swirls past the buttresses of its many bridges …

Hume Nisbet (1849–1923)

After the Thames passes through Eynsham Lock and King's Lock, then around the high ground of the University of Oxford owned Wytham Woods, the river passes below the A34 Oxford Western Ring Road, erected in 1961. Constructed in steel and concrete, the crossing's official name is Thames Bridge. From this point onwards the river runs towards the city of Oxford, where debates continues over whether the river should be known as the Thames or the Isis.

Thames Bridge.

Godstow Bridge.

On reaching Oxford, the river flows below stone-built Godstow Bridge, which marks the northern boundary of the city and university. Constructed in two parts, the oldest section dates from medieval times and crosses the original Thames course, while the newest part, originally erected in 1792 over a lock and cut, was rebuilt in the late 1800s.

It was on this bridge in 1644 where the Royalist army held out against a fierce attack from a force of Parliamentarians during the Civil War. A year later, at nearby Godstow House, a former nunnery, the Parliamentarians burned out the building in a reprisal after the Royalists fled the stronghold and a majority of the remains were later demolished.

Close to Godstow Bridge is the famous Trout Inn, believed to have once been fishermen's cottages. The building dates from the seventeenth-century.

Flowing south towards the city of Oxford, the Thames is partly crossed by Medley Footbridge onto Fiddlers Island, another footbridge carrying the footpath over a cut onto the north bank. To the north of the city, the next crossing is Osney Bridge, which was erected in 1888 to replace a partly collapsed arched stone bridge, believed to have been erected by Osney Abbey monks.

The building of the new cast-iron bridge was overseen by borough engineer of Oxford William Henry White; the castings were made by Horseley Ironworks, Staffordshire; and the abutments, which used stone from the original bridge, were constructed by W. J. McKenzie of Westminster. Osney Bridge has the lowest

Osney Bridge.

headroom of any bridge on the navigable Thames, which currently restricts some boats from passing below.

The main course of the river runs south-eastward past marinas and locks, and on towards the south of the city, where the Osney Railway Bridge spans the Thames, carrying trains on the Cherwell Valley Line. Built in iron, the bridge, which replaced an earlier rail crossing, was completed in 1887 by the Great Western Railway, and is now maintained by Network Rail. On the footpath close to Bulstake Stream, before reaching the railway bridge, there is a tall obelisk erected as a memorial to Edgar Wilson, who drowned on 15 June 1889 after saving two boys who had fallen into the river.

Osney Rail Bridge.

Gasworks Bridge. (Courtesy of DR Neil Clifton)

As the Thames is joined by Castle Mill Stream, the river running on towards central Oxford, its course is crossed by the redundant Gasworks Railway Bridge, linking the district of St Ebbes to Grandpoint Nature Park. Built by the Oxford & District Gas Company in 1886, the bridge was constructed in iron sections then floated into position on barges before assembled on central cast-iron piers sunk into the riverbed. First used by trains carrying coal via a branch line to the gasworks, Oxford's first large-scale industrial site, the Oxford Gas, Light and Coke Company, established in 1818.

After the works closed in 1960 and the buildings were then demolished, the bridge was converted for use as part of a private road, before being used as a pedestrian footbridge.

Another surviving remnant of the gasworks is Grandpoint Bridge, erected in 1927 to carry gas pipes between sites located each side of the river. Named after an

Grandpoint Bridge.

area south of the bridge, Grandpoint was adapted for use as a workers' footbridge and then a public footbridge when the works closed.

Erected on the site of a ford where it was said oxen were once driven across the Isis, Folly Bridge, designed by London Architect Ebenezer Perry and which opened in 1825, is probably the fourth of a succession of bridges to have spanned the river at this point. The first, a wooden structure, was believed to have been built by the Saxons; the second was erected in stone by the Normans.

Through an 1815 Act of Parliament, Folly Bridge, which spans the river in two sections, separated by an island, was erected to replace a bridge that was in a poor state of repair – the navigable arch had fallen into decay and was impassable for rivercraft, and the weir and flash lock below had become a hazard to navigation.

A new lock was added in 1830 and the original tollbooth was rebuilt, which, along with the bridge, are now Grade II listed. Folly Bridge takes its name from a tower 'Welcomes Folly', which stood on the previous crossing known at the time as Southbridge.

Leaving the city centre behind, the river makes up the boundary line between Oxfordshire to the east and Berkshire to the west, spanned by Donnington Bridge, which carries the B4495 road towards Iffley.

Folly Bridge.

Donnington Bridge.

Designed by Oxford engineer J. Campbell Riddell, the bridge was constructed using reinforced concrete and precast concrete-clad abutments, the facia of the bridge was faced with flint, and the abutments faced with decorative Criggion Green and Blue Shap stone.

The bridge was officially opened on 22 October 1962 by Viscount Hailsham, an undergraduate of Christ Church College and MP of Oxford.

To the south of the city, the Oxford ring road passes over the Thames by way of Isis Bridge, on the reach between Sandford Lock and Iffley Lock. Opened in 1965, the single-arch bridge was constructed in steel with a composite concrete slab decking, designed by the British Constructional Steelwork Association. In 2003, the bridge was strengthened and the steelwork received a full wet and dry blast coating.

Further south of the ring road, Kennington Railway Bridge, a steel three-span bowstring structure erected in 1923, carries freight trains on a single-track branch line serving the BMW Mini works at Cowley, a section of the former Wycombe Railway line linking Oxford to Maidenhead.

Designed in-house at Great Western Railway by A. C. Cookson, the bridge was contracted to George Palmer of Neath, and sub-contracted to Horseley Bridge and Engineering Co. Ltd. Each bowstring section was lifted into place by two rail-mounted cranes positioned on the adjacent previous railway bridge, which was then demolished and the track realigned to each end of the new bridge.

From here the course of the Thames runs to Sandford Lock, which has the deepest fall of all locks on the river. Once the site of a mill built by the Knights Templer during the late thirteenth century, the river is crossed by the Thames Path along a weir and onto an island (the original east bank of the Thames before

Kennington Railway Bridge.

Sandford Lock was constructed), with the path then crossing back over Sandford Footbridge to the south.

The footbridge was erected over the original course of the Thames, close to where a ferry once carried horse and rider across the river, marked by a mounting stone – the rider stepping up on the stone to remount the horse.

Sandford Footbridge.

Nuneham Railway Bridge. (Courtesy of DR Neil Clifton)

The river then meanders westward, to where Nuneham Railway Bridge carries the Cherwell Valley line across the Thames. Of similar design to Kennington Railway Bridge, primarily constructed in steel, the single-bow bridge, erected in 1929, was the third in a succession of railway bridges to cross the river at this point. The railway bridge was named after Nuneham House, an eighteenth-century Palladian style villa situated in the local vicinity.

In 1943, the bridge was the starting place for the second unofficial wartime boat race between the universities of Oxford and Cambridge, attended by around 7,000–10,000 spectators; the Oxford team won by two-thirds of a length.

At the small town of Abingdon a crossing over the Thames is made up from three linked bridges. Hart Bridge (also known as Town Bridge) takes its name from the White Hart Inn, Burford Bridge is a corruption of Borough Bridge, and Maud Hales Bridge is named after a wealthy businessman. The bridges span the river with Nag's Head Island situated in between, which is named after the island's Grade II listed public house, the Nag's Head on the Thames.

Constructed using local quarried limestone, the building of the river crossing began in 1416. After it was rebuilt in 1927, the bridges, now known collectively as Abingdon Bridge, consist of a north section with six arches spanning a backwater and millstream, a south section made up of one main arch and four minor arches across the river, and two minor arches on the flood plain.

The construction of the original crossing was funded by the religious guilds of Abingdon, the 300-strong bridge-building workforce supplied with victuals by

Abingdon Bridge.

the good women of the town. Grade II listed, the bridge has been registered as a Scheduled Ancient Monument.

South of Abingdon, the Thames loops eastward towards Sutton Pools, an area on the course of the river consisting of weirs and islands linked by a series of footbridges.

Sutton Pools footbridges.

Sutton Bridge.

The upper level of the Thames is separated from the lower pools by a Saxon causeway and the early nineteenth-century Culham Cut, constructed to take all river traffic past the pools to Culham Lock, the water course rejoining the Thames north of Sutton Bridge.

The stone-built Sutton Bridge, erected in 1807, was originally a private toll bridge that replaced the previous bridge and a ferry. Two years after it was completed, an extension was built across Culham Cut, which opened in 1811.

After the bridge was bought jointly in 1938 by Oxfordshire and Berkshire county councils, the Thames making up the boundary between both counties, the toll was then lifted, and toll house demolished.

Further downstream, the Chellwell Valley line passes across the Thames on the reach between Sutton Bridge and Clifton Lock over the steel-built, bowstring-arch-truss Appleford Railway Bridge. Isambard Kingdom Brunel's wooden bridge, built in 1844 for the Great Western Railway, was replaced in the 1850s by a much more substantial wrought-iron bridge, superseded by the current bridge, which was completed in 1927.

The bridge takes its name from Appleford-on-Thames, to the south, once part of Berkshire up until boundary changes in 1974. The railway line passes through the village where passenger trains stop at Appleford station.

Appleford Railway Bridge.

Originally an Anglo-Saxon settlement, the manor of Appleford belonged to Abingdon Abbey up until surrendering its properties to the Crown during the Dissolution of the Monasteries. At one time Appleford was a major crossing point on the Thames, where apples and cherries from the Harwell orchards were carried across a bridge into Oxfordshire.

Situated on the reach downstream of Clifton Lock, Clifton Hampden road bridge crossed the Thames (up until the boundary changes) from Oxfordshire on the north side, to Berkshire to the south. Opened in 1867, the red-brick-built bridge, designed by Sir Gilbert Scott, well known for Gothic Revival architecture, replaced a ferry that had been in service since the early 1300s.

Clifton Hampden Bridge.

The six-arched, stone-built Clifton Hampden Bridge included five piers with triangular cutwaters extending up to the roadway, to provide a refuge for pedestrians when vehicles were passing. On completion, the bridge builder, Richard Casey, was then appointed toll keeper. The bridge tolls fees were in force up until 1946, after which it became free to cross the Grade II listed bridge.

As the Thames passes Dorchester to the east, turning sharply eastward at Little Wittenham, a footbridge crosses the river in two sections, with Lock House Island in between. Little Wittenham Footbridge, an arch-built, cast-iron and wood structure completed in 1870, crosses the east channel of the river, adjacent to the Days Lock keeper's cottage. The footpath then leads onto a modern built bridge spanning the west channel, leading towards Little Wittenham woods and village.

The arched bridge was one of two bridges once used for the annual Poohsticks World Championships, a game first mentioned by A. A. Milne in his publication *The House on Pooh Corner*, where each player drops a stick on the upstream side of the bridge and the first stick to appear on the downstream side is the winner.

A bridge at Shillingford had existed since the late 1300s, most likely a wood-built footbridge used as part of an access route between Dorchester Abbey to the north and Wallingford Castle to the south. When the porter of Wallingford Castle was granted a ferry for use to carry heavy goods, carts and horses across the Thames, the bridge was then dismantled.

By the mid-eighteenth century over a hundred local landowners formed the Shillingford to Reading Turnpike Trust, created to improve and widen roads,

Little Wittenham Footbridge.

Shillingford Bridge.

and included the building of a new bridge, which on completion made the ferry redundant. Stone piers and abutments were erected to support a wooden trestle toll bridge, which remained in use until replaced by a new stone-built crossing, completed 1827.

The Shillingford Bridge Hotel, situated on the south bank of the river, once known as the Swan Inn, was originally a small dwelling leased by the ferry operator.

Crossing the Thames on the reach between Benson Lock and Cleeve Lock is Wallingford Bridge, the original crossing erected as a means to besiege Wallingford Castle in 1141, during the reign of King Stephen. The stone-built bridge, erected around a hundred years later, was improved and rebuilt during the following centuries.

Part of the bridge was replaced by a drawbridge during the Civil War, which could be raised when the castle came under siege. It was at the market town of Wallingford, to the west of the river, where, in December 1066, William the Conqueror accepted the surrender of Stigand, Anglo-Saxon Archbishop of Canterbury, supporter of King Harold who died at Battle two months earlier.

Wallingford Bridge.

Following a flood in the early 1800s, three of the arches were rebuilt to a design by John Treacher, General Surveyor for the Thames Navigation Commission.

Built as part of a bypass in 1993, during construction of Winterbrook Bridge archaeologists had an opportunity to investigate the remains of a multi-period ancient settlement situated on the south bank of the Thames, as well as an Iron Age earthwork known as Grim's Ditch, which now forms a segment of the Ridgeway National Trail. Designed in a manner that ensured it did not disturb the archaeological site, the three-span bridge was built of steel plate girders and reinforced-concrete deck slabs, and the underside was made up of reinforced-plastic cladding.

The bridge takes its name from nearby Winterbrook, a small settlement where author Agatha Christie lived up until her death in 1976 at the seventeenth-century Winterbrook House.

Known as 'Four Arch Bridge', the current Moulsford Railway Bridge is a combination of two parallel bridges built at skewed angle across the Thames. The first bridge, erected between 1838 and 1840, constructed in red brick with Bath stone quoins, was designed by Isambard Kingdom Brunel and was of a similar design to several other of his bridges on the Great Western Railway line.

The second bridge, erected towards the end of the nineteenth century to accommodate an expansion to the line, followed the design of Brunel's bridge, both connected by girders, brick bridgelets and a series of vaulted brick footbridges. The first bridge is recognised by Historic England as one of the most impressive along the whole length of the rail line.

Above: Winterbrook Bridge.

Below: Moulsford Railway Bridge.

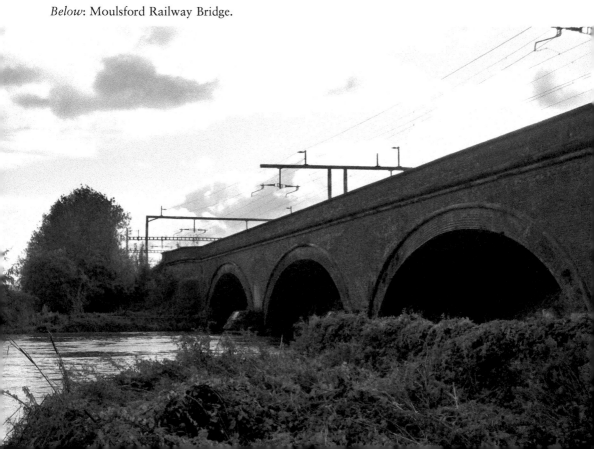

Linking the twin villages of Goring in Oxfordshire and Streatley in Berkshire, the present bridge, named after both settlements and built in 1923, was erected in two parts, with a weir island in between, the rustic timber bridge struts supporting a metal road surface.

The Thames was originally crossed at this point by a Roman-built causeway, and as the river evolved into an important trade route and source of power for water mills, locks and weirs were constructed to control the flow of water.

A toll bridge was then erected between Goring and Streatley in 1837, which replaced a ferry service, where sixty people had lost their lives in 1674 when the ferry capsized in the weir pool. As the twin villages became a popular destination for tourists and day trippers, the new bridge was built to ease traffic flow.

As both villages became more accessible and a fashionable place to visit, the Great Western Railway opened a station at Goring, the trains running over the Thames on the Gatehampton Viaduct, which opened in 1840, another bridge designed by Brunel.

By the end of the nineteenth century, a second rail crossing was erected alongside the first to meet a demand of an increase in rail traffic. Both bridges were built using red brick, laid in English bond, with Bramley Fall gritstone dressings, using a system of internal longitudinal walls and voids to lighten the

Goring and Streatley Bridge.

Gatehampton Viaduct.

superstructure over the four flat semi-elliptical arches – a design Brunel used for many of his railway bridges.

The second surviving private toll bridge over the Thames, on the reach above Mapledurham Lock, crosses the river between the villages of Whitchurch in Oxfordshire and Pangbourne in Berkshire.

There has been a bridge on the site since the late eighteenth century, renewed and rebuilt several times up until the present bridge was reconstructed between 2013 and 2014, the ornate lattice girders of the 1902 bridge incorporated into the new design.

Owned and maintained by the Company of Proprietors of Whitchurch Bridge, tolls to cross the bridge range from 60p for cars up to £4 for heavy vehicles, pedestrians crossing for free. The original charges were a halfpenny each for people and livestock, and 2d for each wheel of a cart or carriage.

Whitchurch Bridge.

Caversham Bridge.

As the Thames flows on towards the historic market town of Reading, the river is crossed during various annual events by the Reading Festival Bridge, linking the festival site on the south bank to a campsite and car park on the north bank.

The temporary bridge, erected on permanent footings, was first used for the Reading Music Festival in 2008, replacing a ferry service that had raised complaints over the long queues waiting to use it. The bridge also eased disruption the ferry caused to passing river traffic.

The first permanent bridge on entering Reading is situated on the reach above Caversham Lock. Caversham Bridge was erected on the site of the first crossing, built during 1163, where a trial by combat was fought out on Fry's Island between Robert de Montford and Henry of Essex, who had been accused of cowardice in battle. Wounded during the contest, Henry was judged guilty as accused, losing his lands along with his honour.

Opened in 1926, by Edward, Prince of Wales, the present Caversham Bridge, an arched concrete structure with granite balustrades, replaced the previous iron lattice crossing after the completion of Reading Bridge in 1923.

Before the Thames reaches Reading Bridge, a pedestrian and cycle bridge links the town centre to a suburb of Caversham, once a place of pilgrimage during the early twelfth century, where people came to visit a shrine dedicated to the Virgin Mary.

The cable-stayed Christchurch Bridge, named after Christchurch Meadows on the north bank, was built to create a safe and easy journey to and from the town, the bridge winning a design commendation in the 2016 Civic Trust Awards. At night the bridge is illuminated with an array of colour changing lights.

Christchurch Bridge.

Reading Bridge, erected in 1923, was the first road bridge built at the site, a half mile upstream from Caversham Bridge, to ease road traffic congestion after the borough boundary expanded into adjoining parishes.

Including the approach road viaduct on the south bank, Reading Bridge, constructed in concrete, Portland stone and steel, is almost 183 metres in length, claimed at the time of construction to be the longest span of its type in the world. The staircases at each end lead down to the old tow path where tunnels allow pedestrians to continue along the riverside paths each side of the Thames.

Reading Bridge.

Sonning Bridge.

At Sonning, the course of the Thames flows below the brick-built arch of Sonning Bridge, the roadway over the bridge extending along two backwater bridges crossing a millrace and Sonning Backwater, with the backwater crossed by a footbridge on the Thames Path.

Erected in 1775 to replace an earlier wooden-built bridge, Sonning Bridge crosses the boundary between Oxfordshire and Berkshire, marked by a stone at the bridge's centre, once the border between Anglo-Saxon Wessex and Mercia.

The Grade II listed bridge has been the subject of many celebrated artists, including J. M. W. Turner. When a proposal was put forward by Oxford County Council to replace the old bridge with a steel and iron structure in 1902, it brought so much opposition from all quarters, as reported in the *Times* newspaper, which rarely devoted much column space to events in the provinces, that the plans were eventually abandoned.

Supported on cast-iron cylinders filled with concrete, the wrought-iron- and cast-iron-built Shiplake Railway Bridge, erected in 1897, carries the Henley branch line across the Thames downstream of Shiplake Lock. The bridge replaced an adjacent timber Great Western Railway bridge, built in 1857 by Brunel's assistant, T. H. Bertram.

After the line was reduced from two tracks to a single track in 1970, the downstream side of the twin-span Shiplake Railway Bridge was removed, leaving just the brick abutments and cast-iron piers in place.

At Henley-on-Thames, a crossing has been in existence since ancient times, the first believed to have been used by the Romans in the first century AD; however,

Shiplake Railway Bridge.

the earliest record of a bridge only dates back to 1232. During the mid-fourteenth century, records also provide evidence there was once two granaries and several chapels situated on a timber and stone bridge, the crossing severely damaged during a great flood in the late 1700s.

The present five-arch, stone Henley Bridge, completed in 1787, was designed by William Hayward of Shrewsbury and erected by John Townsend, a mason from Oxford, at a cost of around £10,000. Henley-on Thames is famed for the Henley Royal Regatta, established in 1851, when Prince Albert became patron of an annual rowing event, which first began in 1839.

Henley Bridge.

Temple Footbridge.

As the Thames Path nears Marlow, it crosses the river by way of Temple Footbridge, erected in 1989, at the site of a former ferry, connecting the counties of Berkshire and Buckinghamshire. The haunched girder rustic-style timber bridge is the longest hardwood bridge in Britain and takes its name from the nearby Temple Mill Island, once owned by the Knights Templar and later occupied by a water mill manufacturing copper sheets used to clad naval warship hulls.

At a height of 6.51 metres above the sea level, the bridge allows passage through by vessels of various height.

The Grade I suspension bridge at Marlow, erected between 1829 and 1832, was built to replace a late eighteenth-century timber bridge further downstream that had fallen into a state of decay. A bridge at Marlow had been in existence since the reign of Edward III, surviving until partly destroyed by Parliamentary forces during the English Civil War.

A new Marlow Bridge was initially designed by John Millington, who then resigned from his position after receiving criticism of his work and project supervision. Bristol-born civil engineer William Tierney Clark was then appointed in Millington's place and given the task of redesigning Marlow Bridge. Clark's bridge design became the prototype for the larger suspension bridge built across the Danube in Budapest.

Marlow Bridge.

After restoration in 1965, a weight limit restriction was then placed upon Marlow Bridge, with vehicles above 3 tonnes diverted from the historic centre of Marlow to a bypass bridge, which carries the A404 between Maidenhead and High Wycombe.

The Marlow bypass bridge was completed in 1972, crossing the Thames between Cookham Lock and Marlow Lock. A balanced cantilever design, constructed in prestressed concrete, an outdoor adventure centre stores its boats and equipment below the bridge on the Thames bank.

Marlow Bypass Bridge.

Bourne End Railway Bridge.

After a northward loop in the Thames, the river is crossed by Bourne End Railway Bridge on the Marlow branch line to Bourne End station, from which the bridge takes its name. The first timber railway bridge, another of Brunel's designs, was later rebuilt in steel, as barges found it difficult to navigate the narrow spans and frequently collided with the upright timber stanchions. The iron box girder lattice truss bridge, completed in 1895, was supported on wider span cylinder pillars, which aided vessels navigating along the river.

In 1992, a footbridge was added, which cantilevered out from the railway bridge to take the Thames Path over the river, and the whole structure was restored and repainted in 2013.

When a bridge was built across the Thames at Cookham in 1839 to replace a succession of ferries, it was not the first bridge erected at the site, as the remains of a Roman-built bridge, known as Camlet Way, were discovered during the nineteenth century at Hedsor Wharf, the bridge gradually falling out of use after the Romans left Britain.

The Cookham Bridge Company invited proposals to design a safe and economic river crossing, deciding upon a design by George Treacher over one submitted by Brunel, which the bridge company considered too expensive. The timber-built toll bridge opened in 1839 but required so much costly maintenance during its twenty-eight years of use, it was eventually replaced in 1867 by a much more durable wrought-iron girder bridge, supported by eight pairs of concrete-filled iron pillars.

Cookham Bridge.

The Cookham Bridge Company continued to charge a fee to cross the bridge up until 1947, when the toll was abolished after the company was acquired by Berkshire County Council.

The footbridge at Taplow, which opened in 2018, is one of the newest bridges erected to span the Thames. The elegant, arched bridge, designed by Michael Knight of Knight Architects to complement Brunel's double-arch Maidenhead Railway Bridge, comprises of triangular steel arches and a composite steel-concrete deck.

The completed bridge was floated upriver by barge and placed into position using hydraulic jacks. The bridge links a new residential development at Taplow to Maidenhead on the opposite bank.

Taplow Bridge.

Many of the river's wooden-built bridges were often destroyed during periods of conflict or fell into disrepair through lack of financial investment, and the estimated cost to repair and restore the last timber-built Maidenhead Bridge was so expensive that the Maidenhead Corporation, charged with maintaining the bridge, concluded it would be more cost-effective to rebuild the crossing altogether.

After the bridge was completed in 1777, the Corporation increased the toll fees to pay for its construction. Although paying a toll to cross over a bridge, or pass below it by rivercraft, had been an acceptable condition of use on a majority of Thames bridges, the tolls of Maidenhead Bridge were considered overly excessive.

After an Act of Parliament was passed legislating public tolls should cease on the last day of October 1903, large crowds gathered on Maidenhead Bridge on the first day of November, removed the toll gates and with great fervour threw them into the river.

The eighteenth-century Grade I listed bridge, constructed of brick and Portland stone, replaced a succession of timber-built bridges. The first was erected in 1280. A chapel was also built at the same time as the bridge, where travellers would pray for safe passage across the river.

When the newly built Maidenhead Railway Bridge opened in 1838, the Great Western Railway board of directors had serious concerns the innovatively

Maidenhead Bridge.

Maidenhead Railway Bridge.

designed wide arches lacked stability and may well collapse under the weight of trains travelling along line. Erected as a single red-brick structure of two arches that crossed the river, buttressed by two smaller land arches at each end, the crossing was another of Brunel's early bridge designs.

When Brunel was instructed by the railway directors to leave the wooden formwork in place to support the brickwork arches, Brunel simply moved the formwork away slightly, giving the appearance the arches were still supported.

During a storm a year later, the wooden supports were washed away, and the bridge has remained standing ever since, justifying Brunel's ingenious design. Heralded as a marvel by contemporary observers, Brunel's bridge was the subject of J. M. W. Turner's atmospheric painting *Rain, Steam and Speed*.

The bridge was widened in the late nineteenth century to carry four tracks, the work carried in such a way to ensure its appearance preserved Brunel's original design.

To the south of Maidenhead, the Thames is divided by a group of low-lying islands at Dorney Reach where the river is crossed by the Thames Bray Bridge carrying the M4.

Erected in the 1960s, the motorway bridge was widened by 8 metres as part of a major building programme during June 2020 to match up with the new

Thames Bray Bridge.

smart motorway system. The extended bridge elements made up of 400 tonnes of steelwork were assembled at a nearby site then lifted into place.

Past the southern tip of Monkey Island, also known as Monks Island, where monks once held fishing rights, Summerleaze Footbridge links the villages of Dorney in Buckinghamshire and Bray in Berkshire.

The bridge, which opened in 1996, had a dual purpose: it was used as a footbridge and cycle track above, and conveyor belt below. The conveyor moved around 4 million tonnes of aggregate from the purpose-built Dorney rowing lake, a venue for the 2012 Olympics, to the Summerleaze processing plant on the opposite side of the Thames, the bridge taking its name from the project contractors.

Summerleaze Footbridge.

3

Windsor to Brentford

See, this regal Thames is winding; among its poplared islands with a slow majestic pace; we should see the towers of Windsor if the sun were not so blinding, it casts a glow on all the trees, and glory on your face ...

Bessie Rayner Belloc (1829–1925)

To the west of the town of Windsor, the A332 dual carriage bypass crosses the Thames on the Queen Elizabeth Bridge, originally named Windsor Way and opened in 1966. One of the first bridges designed by architects rather than by engineers, as was the tradition in the past, the haunch girder bridge supports at each end are held in place by huge chains, which also support the central roadway.

By the early 2000s the bridge needed extensive repair due to corrosion of the metal components, resulting in a £2.1 million makeover to replace twenty tie rods, make concrete repairs, refurbish expansion joints and waterproof the structure. The project was completed by late 2021. To the east of the bridge on the south bank, there is a slipway where boats up to a maximum of 6.7 metres can be launched adjacent to the Windsor Boys' School boathouse. The school, founded in 1908, is one of the top school rowing clubs in the UK, competing in many prestigious boating events, including the Henley Regatta, the boys' school winning their class seven times.

Queen Elizabeth Bridge.

Windsor Railway Bridge.

The building of Windsor Railway Bridge, a single-span, wrought-iron, bow-and-string bridge designed by Brunel, was delayed through objections raised by the provost of Eton College as the bridge encroached on college grounds. Approval was finally granted in 1848, with provision for the protection of the college amenities, and the line was opened a year later. The bridge is thought to be the oldest wrought-iron railway bridge in the world still in regular use.

The line runs along a low, brick-built viaduct, originally constructed in timber, leading up to the Thames from the south, then onto Baths Island, named after the town's first swimming baths, before crossing the river's main course by way of the bridge and onwards across the viaduct on the opposite bank. The viaduct is one of the longest in the world.

The Thames Path runs below the railway bridge on the north bank and then leads westward onto Windsor Bridge, which opened in June 1824. It is believed that a bridge was first erected across the Thames at this site during the twelfth century, where tolls were levied on vessels passing through below.

Windsor Bridge.

Rebuilt in 1242, the oak timber used in the bridge's reconstruction came from trees cut down in Windsor Forest. After a continuation of repairs and refurbishments, by the early 1800s the bridge was in such a poor state of repair it was taken down and replaced by a three-arch, cast-iron and granite-built toll bridge, which linked Windsor to Eton.

The Grade II listed bridge, designed by Charles Hollis and built by William More, completed after his death by his executor, Mr Baldock, was closed to traffic in 1970 after cracks were found in the bridge's supporting cast-iron girders. The damage had been caused by the heavy traffic, which the bridge was never designed to carry.

After repairs were made the bridge was reopened to pedestrians and cyclists only, with road traffic diverted by way of the Queen Elizabeth Bridge to the west, and Victoria Bridge and Albert Bridge to the east.

Before reaching Victoria Bridge, the Thames is crossed by the three-span Black Potts Railway Bridge on the reach above Old Windsor Lock, carrying the line from London's Waterloo station. Expected to be completed in August 1849, when one of the three cylindrical supporting piers began to sink and a cast-iron girder of the bridge snapped, it would be a further three months before the line was officially opened by Queen Victoria after repairs were completed.

Originally, the bridge had ornate cast-iron ribs, but after corrosion they were all replaced with wrought-iron girders, altering the appearance of the railway bridge.

Black Potts Railway Bridge.

Named after Black Potts Ait, an island in the middle of the river supporting the bridge, it is said that Charles II fished close by at a bend in the river while in residence at Windsor Castle.

Downstream, two bridges, one named after Queen Victoria, the other after Prince Albert, were built to replace the timber-and-stone-built Datchet Bridge, demolished in the mid-1800s on completion of both new bridges.

The Victoria Bridge was designed by civil engineer Thomas Page, and it is said that Prince Albert also had a part in the bridge's design, which was constructed in cast iron with stone abutments.

Severely damaged by tanks crossing the Gothic-style bridge while on manoeuvres during the Second World War, a weight restriction was imposed on the bridge until repairs were made in the early 1960s. The ornate cast-iron girders were then replaced with much plainer-designed supports.

Built to a similar design of Victoria Bridge, the first Albert Bridge fell into disrepair due to a dispute between the counties of Berkshire and Buckinghamshire over maintenance and upkeep costs, the bridge linking both counties.

After authorities eventually came to an agreement to fund the bridge's replacement, the old bridge was demolished in 1927 and a new crossing, made up

Victoria Bridge.

of two arched spans constructed in stone and concrete, was erected between Old Windsor and Datchet.

Old Windsor, once the site of a palace of Saxon kings, came into the possession of William I, and the medieval manor house became a royal hunting lodge after Windsor Castle was built. Datchet was also once an Anglo-Saxon manor, later bestowed by William I to Norman baron Giles de Pinkney.

Albert Bridge.

Runnymede Bridge.

First proposed in 1939 to a design by renowned English architect Sir Edwin Lutyens, the Runnymede Bridge was delayed in its construction first by the outbreak of the Second World War, and then by the architect's death in 1944. The design was later adapted by Lutyens' colleague George Stewart, and the single-span bridge, supported by eighteen encased steel arches, was eventually erected across the Thames in the 1960s. Carrying the A30, the bridge was later expanded to carry the M25, as well as a pedestrian and cycle path.

Because of its close location to the historical water meadow where King John sealed Magna Carta in 1215, the bridge was aesthetically finished in handmade brick facings and Portland stone. In 1983, the New Runnymede Bridge was built adjacent to the first, which forms the eastern side of the M25 motorway.

The single-arch bridge, supported by a series of parallel supporting concrete frames, was designed by Ove Arup and Partners to compliment the design of the first bridge, rather than copy it. During excavation work at the site, various artifacts were discovered that date to the Bronze Age, along with evidence of a late Bronze Age riverside settlement.

One of the first Roman-built bridges to cross the Thames was erected at Staines, which lay on the Devil's Highway, a Roman road connecting Londinium

(London) and Calleva Atrebatum (Silchester) via Pontes (Staines). 'Pontes' translates as 'bridges', suggesting there was more than one bridge crossing the river at Staines.

When the Danes were raiding settlements along the Thames Valley, they crossed the Thames over a bridge at Staines to evade Anglo-Saxon forces gathering in London. A bridge was also mentioned in the early thirteenth-century records, listing a tree from Windsor Forest given over for the bridge's repair by Henry III. Refurbished and repaired over several centuries, the wooden bridge was totally destroyed during the English Civil War.

Although later rebuilt in wood, the bridge was replaced in the late 1700s by a stone-built structure designed by Thomas Sandby, Professor of Architecture at the Royal Academy. This bridge, however, partially collapsed soon after its opening in 1791, replaced by a succession of cast-iron bridges.

The current crossing was designed by George Rennie and John Rennie the Younger, and was built in smooth, rustic granite with brick approaches. The toll bridge was opened by William IV in 1832.

The U-shaped iron girder railway bridge at Staines, built in 1856, passes over the Thames to the south-east of the town centre, built to carry an independent

Staines Bridge.

Staines Railway Bridge.

M3 Chertsey Bridge.

railway company line from Staines to Wokingham. Taken over soon after by the London & South Western Railway, the line is on the through route from Waterloo station to Reading.

The Thames Path passes below the railway bridge to the north side of the river, leading towards the concrete-built M3 motorway bridge, erected in the early 1970s, and then to Chertsey Road Bridge, 460 metres downstream.

The Grade II listed Chertsey Bridge, erected in the late 1700s to replace a previous crossing, was constructed in Purbeck stone and consisted of seven arches, each formed of a segment of a circle. Paid for by the counties of Surrey and Middlesex, the crossing linking the two, the bridge was built under direction of the architect James Paine at a cost of £6,813 4s 11d.

On nearing completion, it became evident the bridge would not reach the Surrey bank, and the county was required to supply the funds to complete the extension work required.

The first bridge at Chertsey, built of timber in the late thirteenth century, was erected close to where a ferry had once transported goods, livestock and people across the Thames, including Edward I and his family when travelling to Kingston in 1299; the royal party were charged a toll fee of 3s to cross. Kingston has been associated with royalty since the first century where Edward the Elder was crowned King of the Anglo-Saxons in AD 900.

When the first bridge at Walton-on-Thames was erected in 1750, it was built close to the site of several barrows where various earthenware vessels and spearheads were discovered during excavations.

Chertsey Bridge.

Fifth Walton Bridge.

The ornate, mathematically designed bridge was considered an impressive feat of engineering at the time, where a complex arrangement of straight lengths of timber, fixed into place by joists, formed an arch. Depicted in a painting by Canaletto, the bridge lasted less than forty years, and after falling into decay the timber bridge was replaced by a stone-clad brick bridge.

The latest crossing, a thrust-arch, single-span road bridge erected at a cost of £32 million, opened in July 2013, the sixth in a succession of crossings on this stretch of the river at Walton.

As the Thames flows eastwards it is crossed by Hampton Court Bridge, the first road crossing on the boundary of Greater London, and the only London bridge leading to another county. From Surrey to the south, the A309 runs over the bridge towards the magnificent Hampton Court Palace, a royal residence of kings and queens since the reign of Henry VIII. Prior to the building of the first bridge in 1753, a ferry had been in operation since the late fifteenth century, moving people and goods across the Thames to and from the palace.

The fourth in a succession of bridges, Hampton Court Bridge, designed by architect Sir Edwin Lutyens and Surrey County engineer W. P. Robinson, was built in a style to reflect sections of the palace designed by Sir Christopher Wren.

Hampton Court Bridge.

Opened in July 1933, by Edward, the Prince of Wales, the crossing was Grade II listed in 1952.

Downstream from Hampton Court, a succession of bridges had existed at Kingstone-Upon-Thames since at least the Anglo-Saxon period, although situated further down the river from the current bridge. Kingston takes its name from the King's Estate and the stone used as a seat for the coronation of the kings, recorded in the early 800s, as Cyninges-tun.

The first recorded bridge was erected sometime towards the late thirteenth century, on the ancient border between Wessex and Mercia. After falling into disrepair, the bridge was replaced by a succession of toll bridges, one of which was severely damaged during the Wars of the Roses.

In 1825, an Act of Parliament was passed that authorised the building of a new bridge upstream, designed by county surveyor Edward Lapidge to replace the previous crossing, which had partially collapsed during a severe frost.

Kingston Bridge.

The five-arch bridge, constructed in Portland stone, was opened to traffic in 1828, and, after lengthy negotiations, made free from tolls in 1870. In the early 1900s, the bridge was widened on the downstream side to accommodate an increase in traffic and for use by trams.

Further downstream from the road bridge, Kingston Railway Bridge crossed the Thames carrying a line linking Kingston and Hampton Wick, part of a loop line running out of Waterloo station.

The first railway crossing, a deck arch bridge constructed in cast iron, which opened in 1860, was replaced in 1907 by the current railway bridge designed by John Wykeham Jacomb-Hood, chief engineer of the London & South Western Railway. Historically within the county of Surrey, Kingston was absorbed into Greater London in 1965.

Kingston Railway Bridge.

Erected between 1887 and 1889, the two footbridges crossing the Thames at Teddington, coming together on an island, replaced a ferry service south of the lock. The east bridge from Ham is a simple iron girder design, whereas the west crossing is a more striking suspension bridge, leading across the weir stream to the town of Teddington.

From each end of both footbridges, the Thames Path runs downstream on both banks of the river. An obelisk on the Surry towpath marks the point from where the Port of London Authority has control of Thames navigation downstream all the way to the North Sea.

Right: Teddington Suspension Footbridge.

Below: Teddington Lock and Girder Footbridge.

Richmond Bridge was commissioned in the early eighteenth century to replace a ferry that had been in existence since the Norman period, and was later used to serve Richmond Palace, built for Henry VII, and named after his Yorkshire earldom of Richmond. Its construction brought some hostility from the locals and objections were made against its construction, including fears the abutments, unprovided with arches, would cause flooding in the general area at high water.

After the majority of grievances were resolved and the ferryman was financially compensated for the loss of his ferry service and earnings, architects James Paine and Kenton Couse were commissioned to design the new bridge. However, further objections were raised over the position of the crossing and the Bridge Commission was forced to erect the bridge at the old ferry site, the high bank each side of the river making this a much less suitable location.

Funded through the sale of shares, the building of the elegant five-arch, stone toll bridge began in 1774, but due to delays was not completed until December 1777. Although delayed, the bridge became a financial success, generating a profitable return for shareholders through the toll charges.

As the course of the Thames flows northwards, it is crossed again by the loop line from Kingston on the Richmond Railway Bridge, rebuilt between 1906 and

Richmond Bridge.

1908 as an almost exact replica of the original cast-iron railway bridge that first came into service in 1848.

After a bridge of similar design collapsed at Norbury Junction during the late 1800s, the original Richmond Railway Bridge was dismantled and rebuilt by the Horseley Bridge Company, to a design by John Wykeham Jacomb-Hood.

The original piers and abutments were retained, and the cast-iron girders were replaced with steel girders. Further refurbishments were carried out during the mid-1900s, replacing the decking and main girders, and to protect the historic structure Richmond Railway Bridge was Grade II listed in 2008.

Running adjacent to the railway bridge, Twickenham Road Bridge, opened by Edward, Prince of Wales, in 1933, is a section of an arterial road connecting the suburb of St Margarets on the north bank to the district of Richmond on the south bank, close to Richmond Palace, onetime residence of Henry VII, adjacent to the Old Deer Park, an area of open land owned by the Crown Estates.

The crossing, designed by architect Maxwell Ayrton, famous for designing the old Wembley Stadium, takes its name from nearby Twickenham, recorded in the early AD 700s as Tuiccianham – the name is also associated with the word *twicce*, meaning 'river fork'. The bridge's three reinforced-concrete arches, supported

Richmond Railway Bridge.

Twickenham Bridge.

on concrete river piers, incorporate bronze plate hinges at their base and centre, allowing the structure to adjust with temperature changes – the first bridge in the UK to use such a system.

As the Thames flows downstream, around the celebrated Eel Pie Island, named after the eighteenth-century Eel Pye House, a one-time eminent place of entertainment where eel pies were served to merrymakers, the tidal water of the river comes under the control of Richmond Lock and Weir. The weir is topped by twin footbridges running across five high, ornate metal arches, supported on large brick piers protected by ashlar stone cutwaters.

Designed in a collaboration between surveyors and engineers, the conventional lock system was capable of handling up to six river barges, the weir controlled by lifting and lowering sluice gates cranked by hand.

The footbridge was officially opened by the Duke and Duchess of York in May 1894. Pedestrians paid a toll to cross the footbridge up until 1938, when the toll was abolished.

The only lock on the Thames owned and operated by the Port of London Authority, Richmond Lock underwent a major £4 million refurbishment during the 1990s.

Richmond Lock and Footbridge.

4

Kew to the Mouth of the Thames

Twenty bridges from Tower to Kew (Twenty bridges or twenty-two); wanted to know what the River knew; For they were young, and the River knew; for they were young, and the Thames was old; and this is the tale that River told …
Rudyard Kipling (1865–1936)

After a majority of bridge tolls were abolished during the mid-nineteenth century, in 1873 the gates of Kew Bridge were taken down and ceremonially paraded around Kew Green before a celebratory firework display took place in the evening. Soon after, however, it became clear that Kew Bridge, the second crossing built at the site, could not cope with the extra weight of traffic using the free crossing, becoming regularly congested at the narrow Brentford approach road end.

Towards the end of the nineteenth century, a new wider crossing was commissioned by Surrey and Middlesex county councils, and a temporary timber bridge was erected upstream while the old bridge was demolished and the new one built. This third bridge at the site, officially named King Edward VII Bridge, was opened in May 1903 by Edward VII, after His Royal Highness laid the last coping stone.

Kew Bridge.

A grand celebratory party then took place on the bridge, which later overflowed onto the lawns of Kew Gardens. The area of Kew is believed to have taken its name from a key-shaped piece of land on the Surrey bank, recorded in 1327, as Cayho. The rough granite-built bridge arches are decorated with the shields of the counties of Surrey and Middlesex.

To the west of the road bridge, Kew Railway Bridge, crossing the Thames from Kew to Chiswick, was opened in 1869, carrying the line from Acton Junction to Richmond. Designed by William Robert Galbraith for the London & South Western Railway, the wrought-iron lattice girder bridge is supported on four pairs of cast-iron piers.

The embankment ends each have red-brick-built, stone-dressed, arched entrances below track level. The overhead line is now used by London Overground and London Underground train services.

To relieve traffic congestion west of London, a new bridge was built across the Thames to carry the A316 road between Mortlake on the south bank, and Chiswick on the north bank, the name of Chiswick meaning 'a cheese farm'.

Chiswick Bridge, a reinforced-concrete deck, three-arch bridge faced with Portland stone was open to traffic in 1933, erected on the site of a former ferry crossing. Designed by architect Sir Herbert Baker and engineer Alfred Dryland, at the time of building Chiswick Bridge had the longest concrete span over the river.

Kew Railway Bridge.

Chiswick Bridge.

Now on a major transport route around the west of London, Chiswick Bridge is at one end of the course for the universities of Oxford and Cambridge Boat Race, marked by a stone on the south bank.

During the late nineteenth century, after the collapse of the railway bridge at Norbury Junction, there were concerns about the suitability of cast-iron-built bridges carrying heavy locomotives and loads, and although the cast-iron Barnes Railway Bridge showed no signs of failing, it was decided to build a new wider bridge alongside the first.

Designed by Edward Andrews, the new bridge, named after Barnes, a riverside settlement first recorded in the Domesday Book as 'Berne', comprised of three

Barnes Railway Bridge.

spans of wrought-iron bowstring girders, carrying two tracks and a pedestrian walkway. After the new Barnes Railway Bridge opened in 1895, the original bridge was no longer used, and there are currently plans to convert the Grade II listed structure into a green walkway across the Thames.

The ornate Hammersmith Suspension Bridge, designed by civil engineer Sir Joseph Bazalgetti, which opened in 1887, replaced the first ever suspension bridge built across the Thames in London.

After a boat collided with the original bridge in 1882, the new bridge was commissioned through an Act of Parliament. The new wrought-iron bridge, supported on the piers of the previous crossing, was named after Hammersmith, believed to have been an important area for metal workings, hammers and smithies. Hammersmith Bridge has been targeted on three occasions by IRA bombers during its lifetime, and in 2000 the bridge had to be closed to traffic for extensive repairs.

After reopening, although the Grade II Hammersmith Bridge carried a vehicle weight restriction, it continues to suffer from structural problems brought about by the high volume of heavy traffic using the bridge, which it was never designed to carry.

Hammersmith Bridge.

The Cornish granite, stone and concrete Putney Bridge, also designed by Sir Joseph Bazalgetti, was opened by the Prince and Princess of Wales in 1886, at the site of a eighteenth-century wooden toll bridge, the first crossing to be built over the tidal Thames between Kingston Bridge and London Bridge.

The current Putney Bridge, widened in 1933 to carry three lanes of traffic, now carries four lanes, which includes a bus and cycle lane. Linking the parishes of Putney in the south and Fulham in the north, it is the only bridge in Britain that has a church at each end – St Mary's at Putney and All Saints at Fulham.

On Putney Embankment, below Putney Bridge, a stone marks the start of the University Boat Race Championship Course, first used in 1845, and to the west of Putney Pier are a host of some of London's top rowing clubs.

Putney Bridge.

Fulham Railway Bridge, originally given the name Putney Railway Bridge by its design engineers William Jacob, assistant to Brunel, and W. H. Thomas, is made up from five wrought-iron lattice girders supported by pairs of cast-iron caissons sunk into the clay riverbed.

Built by North Yorkshire-based heavy industrial company Head Wrightson, the railway bridge, which opened in 1889, carried trains on the London & South Western Railway, and later London Underground trains over the Thames on the District Line. Refurbished in the late 1900s at a cost of £8 million, the bridge also carries a pedestrian walkway on the downstream side.

The railway bridge takes its name from the ancient parish of Fulham. The word Fulham originates from Old English: Fulla, the name of an Anglo-Saxon landholder and '*hamm*', meaning an enclosed area.

The Wandsworth Toll Bridge, erected across the Thames between Wandsworth and Battersea in 1873, became a commercial failure as it was expected there would be an increase in traffic crossing the Thames to the proposed west terminus of the Hammersmith and City Railway on the north bank; however, the development didn't go ahead.

Too narrow and not strong enough to carry modern transport vehicles of the early 1900s, the toll fees were abolished when the bridge was taken into public ownership. The bridge was then demolished and a new steel cantilever bridge,

Fulham Railway Bridge.

Wandsworth Bridge.

designed by London County Council chief engineer Sir Thomas Pierson Frank, was erected in its place.

Opened to traffic in 1940, shortly after the outbreak of the Second World War, the bridge was painted in shades of dull blue as camouflage against German air attack. Wandsworth takes its name from the River Wandle, a tributary flowing out into the Thames just upstream from Wandsworth Bridge.

The railway bridge at Battersea, which opened in 1863, was built to serve as London's only north to south through route for railway traffic. Designed by William Baker, chief engineer of the London & North Western Railway, the bridge, formed by five iron-built arches supported by stone-faced piers, crossed the Thames between Chelsea Harbour to the north, now an area of regeneration consisting of a design centre, boat harbour, hotel and luxury apartments, and Battersea to the south.

During the Anglo-Saxon period, the Thames was much shallower and wider at this point than it is today, where there were many small islands along its course. One island reclaimed for agricultural use, known in Old English as Badrices īeg, meaning Baldric's Island, gradually evolved into the settlement of Battersea.

Battersea Railway Bridge.

After refurbishment during the late 1960s, two arches at the southern end of Battersea Railway Bridge were used as a storage area by a houseboat community moored downstream, while an arch on the north side was opened to provide a route through for the Thames Path.

To the east, past Chelsea Creek, another of London's many tributaries, Battersea Road Bridge was opened in 1890, by future prime minister and chairman of London County Council, Lord Rosebery. Consisting of five segmental spans, each span formed by seven cast-iron arches supported on granite piers, the bridge was another of Sir Joseph Bazalgette's designs.

The bridge replaced an earlier timber toll bridge erected during the late eighteenth century at the site of Battersea Ferry. River traffic on the Thames has collided with the bridge piers on various occasions. In one incident in 2005 a 200-ton barge collided with the bridge, causing severe structural damage; the

Battersea Bridge.

crossing was closed for four months while repairs were carried out – a cost of over £500,000.

In 2006, a northern bottlenose whale was found lost in the river close to Battersea Bridge, but despite rescue attempts to save it the whale died. Although it is rare to see a whale this far up the Thames, whales and dolphins are occasionally spotted further downstream.

The Albert Bridge Company, formed in 1863, raised an initial £55,000 in a share capital with the aim of building a new bridge to cross the Thames east of the old, dilapidated, timber Battersea Bridge.

Designed by Rowland Mason Ordish, leading engineer who worked on the Royal Albert Hall, the cantilever toll bridge consisted of four ornamental cast-iron towers, erected on concrete foundations, which supported the carriageway by cable-stayed rods, fanning out from the top of the towers, and a series of suspension chains.

Although named after Albert, Prince Consort, the bridge acquired the nickname 'Trembling Lady', due to the vibration caused by traffic using the bridge and when

Albert Bridge.

troops from Chelsea Barracks were marching across, which led to warning notices posted informing the soldiers to break step for fear the rhythm of troops marching in unison would cause structural damage.

Even though the bridge was significantly upgraded by Bazalgette in 1884, a weight limit of 5 tons was placed on vehicles using the bridge, later reduced to 2 tons. In the mid-1900s, proposals were made to demolish the bridge, which resulted in robust protests from local residents and the Albert Bridge was saved.

Built along similar lines to Hammersmith Suspension Bridge, when work began on the first Chelsea Suspension Bridge in 1851 it was initially named the Victoria Bridge, after reigning monarch Queen Victoria.

Strengthened with extra chains in 1880 due to the amount of use by heavy traffic, the narrow bridge was becoming structurally unsound, and the crossing was renamed Chelsea Bridge to avoid any association with royalty if the bridge ever collapsed.

The area to the north of the bridge was once known as Chelchith, from which Chelsea takes its name, and during bridge excavations various Roman and Celtic artifacts were found, including the famous bronze and enamel Battersea Shield. In the early 1900s, through an increase in use by traffic, the bridge was no longer fit for purpose and was replaced by the first self-anchored suspension bridge built in Britain.

Chelsea Bridge.

Designed by architects G. Topham Forest and E. F. Wheeler and engineers Rendel, Palmer and Triton, the new Chelsea Bridge was constructed by Holloway Bros (London) Ltd, with materials sourced from within the British empire. Opened in 1937, by Canadian prime minister Mackenzie King, the bridge became a favourite meeting place for motorcyclists during the early 1950s, racing each other across the bridge at night.

A short distance downstream is Victoria Railway Bridge, also known as Grosvenor Bridge, named after the Grosvenor family who once owned land where Victoria station was built. The bridge was designed by engineer Sir John Fowler and built between 1859 and 1860, carrying two tracks to Victoria station. After widening in 1865, and again in 1907, to increase the number of rail lines, the crossing was extensively rebuilt in steel between 1963 and 1967.

The new four-arch-span crossing consisted of ten bridges, each carrying a single track, joined together as one, supported on the original piers encased in concrete. Designed by Freeman Fox & Partners, the project engineered by chief civil engineer of British Rail Southern Region, H. Cantrell. Victoria Railway Bridge became the busiest railway bridge in the world.

The five-arched, steel-built Vauxhall Bridge – the name Vauxhall derived from the thirteenth-century Falkes' Hall, a manor owned by Anglo-Norman knight Falkes de Breauté – was designed by civil engineer Sir Maurice Fitzmaurice and resident London County Council engineer Sir Alexander Binnie. Opened in 1906 by George, Prince of Wales, the bridge was the first in London to carry trams across the Thames.

Victoria Railway Bridge – Grosvenor Bridge.

Vauxhall Bridge.

The first bridge built at the site replaced Huntley Ferry, the bridge tolls set high in anticipation that the surrounding areas would become a prosperous suburb. After tolls were lifted when the bridges between Hammersmith and Waterloo were purchased by the Metropolitan Board of Works, Parliament then granted permission to rebuild Vauxhall Bridge, funded from rates paid by all London residents.

Architect Richard Norman Shaw was consulted in providing decorative elements for the new bridge and it was decided to erect bronze statues above the piers, designed by architectural sculptures Alfred Drury and Frederick Pomeroy. The statues on the downstream piers depict Science, Fine Arts, Local Government and Education, and on the upstream piers Agriculture, Architecture, Engineering and Pottery.

Situated at the site of a former horse ferry, where people have crossed the river since at least the mid-fourteenth century, the first Lambeth Bridge, which opened in 1862, was rarely used by horse-drawn vehicles due to the steepness of the approach roads, and the crossing soon fell into disrepair.

Closed to traffic in 1910, a new five-span, steel-arch bridge was commissioned to replace the previous bridge. Designed by engineer Sir George Humphreys and architects Sir Reginald Blomfield and George Topham Forrest, the bridge was built by Dorman Long & Co. of Middlesbrough. On completion in 1932, the new Lambeth Bridge was opened by George V.

Sections of the Lambeth Bridge were painted red to match seats in the House of Lords at the nearby Palace of Westminster, and crests displayed on the granite

Lambeth Bridge.

piers each side of the bridge were erected in honour of London County Council, responsible for funding the bridge's construction.

At the southern end of the bridge is the mid-fifteenth-century Lambeth Palace, London residence of the Archbishop of Canterbury. The name Lambeth is believed to have been shortened from the eleventh-century word *lambehitha*, meaning 'a landing place for lambs'.

Up until the mid-eighteenth century, the closest crossing to London Bridge was at Putney, then due to an increasing population and trading activities expanding out from the city, it became necessary to build a crossing to alleviate the congestion on London Bridge.

Although a bridge across the Thames at Westminster had first been proposed in 1664, this had been opposed to by the Corporation of London, and Thames Waterman, concerned a bridge would impact on their trade of moving goods and people across the Thames by boat, barge and lighter. Although opposition continued into the early 1700s, after the building of Putney Bridge, Parliament finally gave approval and a stone-built bridge was erected across the Thames at Westminster, which opened to traffic in 1750.

Westminster Bridge.

Subsiding badly by the mid-nineteenth century, although remedial work was carried out to stabilise the bridge, work began on building a new Westminster Bridge in 1854, designed by architect Charles Barry and engineer Thomas Page. New granite piers were built to support seven elliptical cast and wrought-iron arches of the bridge, crossing the river from the Palace of Westminster on the north bank towards St Thomas' Hospital on the south bank.

Predominantly painted green, the colour of the House of Commons seats, Westminster Bridge is now the oldest bridge in central London.

Erected at the site of the old Hungerford Market on the north bank of the river, the first Hungerford suspension footbridge, designed by Brunel and opened in 1845, was purchased by South Eastern Railways in 1859 to extend the line to the newly built Charing Cross station.

Above:
Hungerford
Railway Bridge.

Right: Golden
Jubilee Bridges.

Waterloo Bridge.

A new bridge, designed by Sir John Hawkshaw, President of the Institute of Civil Engineers, was built to replace Brunel's railway bridge, comprising of nine wrought-iron lattice girder spans to carry the tracks, with a pedestrian walkway on each side. The new crossing was completed and opened in 1864.

After the railway was widened, which dispensed with the upstream walkway, new footbridges were later erected each side of the railway bridge, named the Golden Jubilee Bridges in celebration of the 50th anniversary of the accension of Elizabeth II to the throne in 2002.

After the first Waterloo Toll Bridge opened in 1817, designed by Scottish engineer John Rennie, the bridge named after the Battle of Waterloo, fought in Belgium two years earlier, it gained a reputation as a place where people went to commit suicide by jumping off into the river – an average of thirty attempts made annually during the mid-1800s.[2]

When London Bridge was rebuilt in the 1830s, causing a change in the currents and flow of the water, the sediments around the bases of Waterloo Bridge's piers were gradually being washed away, and although works were undertaken to protect the foundations, subsidence left the crossing in a dangerous condition, and in 1924 the bridge was forced to close.

2. Charles Mackay, *The Thames and its Tributaries* (1840).

Blackfriars Bridge and the first Blackfriars Railway Bridge.

A temporary bridge was then erected while the old bridge was demolished, and the new Waterloo Bridge was built during the Second World War. Designed by Giles Gilbert Scott and engineers Rendell, Palmer and Triton, the steel, concrete and Portland stone-clad bridge was mostly built by female labourers as a majority of the male workforce had been called up to serve, with the bridge gaining the nickname the 'Ladies Bridge'. Granite stone blocks from the original bridge were presented to members of the British Commonwealth to further historic links between nations.

Taking its name from an area north of the Thames associated with the thirteenth-century Black Friars Priory, Blackfriars Bridge, an Italian-style semi-elliptical arched toll bridge clad in Portland stone, was London's third bridge to be built across the Thames.

After completion in 1769, Blackfriars Bridge required various ongoing repairs, and a hundred years after it was built it was decided to demolish the old crossing when plans were made to create the Thames Embankment. Designed by civil engineer Joseph Cubitt, the five-arch, wrought-iron Blackfriars Bridge, with ends shaped like pulpits in reference to the Black Friars, was opened by Queen Victoria in 1869, and a statue of the queen was erected on the north side of the bridge.

Blackfriars Bridge is maintained by Bridge House Estates, a charitable trust first formed by royal charter in 1282 to pay for the upkeep of London Bridge through toll fees and donations. Bridge House Estates, which owns a significant number of properties in the City of London, also cares solely for the upkeep of five of London's road bridges.

Blackfriars Railway Bridge.

Adjacent to the road bridge at Blackfriars is Blackfriars Railway Bridge, the second railway bridge to cross the Thames at this point, originally named St Paul's Bridge after the mainline station St Paul's.

The first railway bridge designed Joseph Cubitt, which opened in 1864, was made virtually redundant within twenty years, after the rapid advancements in steam locomotion made trains much too heavy to travel safely across the bridge.

Demolished in 1886, leaving only the supporting columns of the bridge in place, the new five-span, wrought-iron railway bridge was erected alongside, designed by engineers John Wolfe Barry and Henry Marc Brunel, son of Isambard Kingdom Brunel.

When St Paul's station changed its name to Blackfriars in 1937, the railway bridge was renamed Blackfriars at the same time. At the beginning of the 2000s, Blackfriars station platforms were extended and enclosed across the length of the bridge as part of the £6 billion Thameslink Programme.[3]

The most recent bridge erected across the Thames in London, the Millennium Footbridge, had an unstable beginning after its opening at the beginning of the twenty-first century.

The design of the footbridge was awarded to Sir Anthony Caro and Arup Group, Foster and Partners, after winning a competition organised by Southwark Council and the Royal Institute of British Architects. Located to the east of Blackfriars,

---

3. Fjdconsulting.co.uk/projects/the-thameslink-programme

London Millennium Footbridge.

building work on the flat steel suspension bridge began late in 1998, and when opened in June 2000 was two months behind schedule.

Within two days, however, the bridge had to close due to unanticipated lateral vibration caused by the high number of people walking across. Although the problem was solved by installing varied systems of dampers, the work putting the bridge out of action until early 2002, the crossing is still referred to by Londoners as 'the Wobbly Bridge'.

Southwark Bridge.

On 24 March 1819, the first bridge built at Southwark, another of John Rennie the Elder's bridge designs, was ceremonially opened to the public at midnight. Consisting of three large cast-iron arch spans, at 240 feet in length the central cast-iron span, cast at the Yorkshire foundry of J. Walker & Co., was the longest ever made.

Suffering from similar sediment erosion as Waterloo Bridge, Southwark Bridge was rebuilt in 1921 by Sir William Arrol & Co. to a design by architect Sir Ernest George. Comprising of five steel arch spans, the bridge's four stone supporting piers were positioned to line up with supporting piers of both Blackfriars Bridge to the west and London Bridge to the east, as an aid for vessels navigating their way along the river.

The bridge takes its name from a district to the south of the river, Southwark, the oldest part of south London, originally an Anglo-Saxon settlement evolving into an area of medieval entertainment. Shakespeare's playing company, the Lord Chamberlain's Men, built the Globe theatre at Southwark in 1599, close to where the reconstruction of the Globe stands today.

To serve Cannon Street Railway Terminus, built in 1866, a wrought-iron plate girder railway bridge, supported by cast-iron Doric concrete-filled pillars was erected at the same time as the station. The bridge carried five sets of railway tracks across the river from the south to the financial district of London.

Cannon Street Railway Bridge.

London Bridge – dismantled in 1967.

Designed by Sir John Hawkshaw, South Eastern Railway engineer, the bridge was first named Alexandra Bridge in honour of Princess Alexandra of Denmark, wife of Edward, Prince of Wales. When the bridge was widened in 1893 to carry ten rail lines and strengthened to accommodate heavier locomotives, the walkways each side were removed – one used by rail workers, the other a public toll path.

Rebuilt twice since the first railway bridge was erected, when British Rail carried out extensive renovations in 1982 many of the original decorative features were removed and never replaced. The area of Cannon Street originally known during the seventeenth century as Candlewrichstrete ('the street of candle makers') falls within one of the City of London's ancient subdivisions, still known today as the Ward of Candlewick.

Up until the early 1700s London Bridge was the only way of crossing over the Thames into London. A bridge of some form had existed near this site since the early Roman period, the first crossing believed to have been a row of boats lashed together with a boarded gangway on top.

The current London Bridge, built between 1967 and 1972, was designed by architect Lord Holford and consulting engineers Mott, Hay and Anderson, and the bridge was constructed by John Mowlem & Co., funded by Bridge House Estates. Comprising of three spans of prestressed concrete box girders, the bridge

Tower Bridge. (Alan Armstrong)

replaced the early 1830s stone London Bridge, designed by Joseph Rennie the Elder, which was taken down in pieces after the new bridge was erected alongside, sold to American entrepreneur Robert P. McCulloch, and rebuilt at Lake Havasu City, Arizona, as a tourist attraction.

Rennie's bridge superseded the legendary old London Bridge, dating back to the early thirteenth century, where buildings had once lined its length on each side, and where heads of traitors, including William Wallace, Jack Cade, Thomas More and Thomas Cromwell, were displayed impaled on spikes above the bridge's entry towers.

After the old London Bridge was demolished, reclaimed parts of the bridge, stone balustrades, pedestrian alcoves and archways were reused on other buildings or as decorative features in recreational spaces.

One of the most recognisable crossings over the Thames, Tower Bridge, which appears to be much older than it is, was built towards the end of the nineteenth century as a route to move people and goods to and from the emerging commercial district and docklands in the East End of London.

Towards the late 1800s, tall-masted sailing ships were still bringing cargo up the Thames to the wharfs and jetties of the Pool of London; therefore, a traditional, fixed, street-level bridge across the river would not have allowed ships to pass below. Subsequently, a Special Bridge Committee was formed to resolve the problem.

Although over fifty bridge designs were submitted, none were considered suitable. The committee eventually approved the building of a bascule bridge in 1885, designed by architect Sir Horace Jones and engineer John Wolfe Barry.

Built with two steel, stone-clad, bridge towers supported on stone piers, the towers were connected at the upper level by two horizontal walkways. To allow

Queen Elizabeth II Bridge.

tall ships to pass through, the road level central span split into two equal length bascules, each weighing 1,000 tons, raised by steam-driven hydraulic engines.

Designed in the neo-Gothic style to complement the Tower of London situated on the Thames north bank, Tower Bridge was officially opened by the Prince and Princess of Wales on 30 June 1894.

The furthest road crossing downstream on the River Thames is the 2-mile-long cable-stayed Queen Elizabeth II toll bridge, which opened in 1991, built to ease the increasing traffic congestion at the two adjacent Thames road tunnels. Financed through private funding, Trafalgar House won the contract to build the new crossing. Designed by German civil engineer Hellmut Homberg, the bridge was built as a joint venture between Norwegian construction and engineering company Kværner, Darlington-based Cleveland Bridge & Engineering Company, and the Cementation Company, a large British construction business.

The roadway bridge deck, suspended just over 54 metres above the Thames, allowing all but the tallest sailing vessels to pass below, is supported by fifty-six pairs of cables fixed to two pairs of steel and concrete masts, 84 meters tall, mounted on concrete piers below.

Opened by Elizabeth II, who the bridge is named after, it was expected tolls would be removed once the fees had paid the cost of building the bridge, estimated to be around £120 million; however, toll fees continued after Transport Act

Thames Estuary leading towards the river's mouth.

legislation permitted the introduction of payment schemes on trunk road bridges and tunnels at least 600 metres long.[4]

Although the Queen Elizabeth II Bridge was the last to be erected across the River Thames, east of Tower Bridge, various plans have been proposed to erect other road bridges between the two, including the Canary Wharf Bridge at Rotherhithe and the Riverside Overground Extension at Barking.

Other projected crossings to the west of Tower Bridge included the Garden Bridge between Blackfriars and Waterloo, which fell out of favour due to costs in 2017; the Diamond Jubilee Bridge, another pedestrian and cycle crossing between Battersea and Imperial Wharf; and a pedestrian and cycle bridge between Pimlico and Nine Elms.

All these bridges were proposed to accommodate a projected increase of public and vehicle traffic moving across the river, and although, at the time of writing, none of these planned bridges have come to fruition, undoubtably, as time passes, more bridges will eventually be built to span the ancient and long winding River Thames.

---

4. Legislation.gov.uk/uksi/2013/2249/made

Bibliography

Cove-Smith, Chris, *The River Thames Book* (Imray, 2017)

Duncan, Andrew, *Secret London* (New Holland 1995)

Evens, H. A, *Highways and Byways of Oxford and the Cotswolds* (Macmillan, 1905)

Hutton, Edward, *Highways and Byways of Wiltshire* (Macmillan, 1917)

Jerrold, Walter, *Highways and Byways of Middlesex* (Macmillan, 1909)

Weinreb, Ben and Christopher Hibbert (eds), *The London Encyclopaedia* (Pan MacMillan, 1983)

Websites

Britannica

British History Online

Historic England

Londonist

Old Maps Online

Ordnance Survey

Port of London Authority

River Thames

Thames-Path

The River Thames

Victorian Web

Vision of Britain

Visit Thames

Acknowledgements

The author would like to thank the following for their assistance in research and for information used in compiling this publication: the British Library, Gloucestershire Archives, the National Archives, Oxfordshire History Centre, Richmond upon Thames Local Studies Library and Archive, and Thameside Marina. The illustrations on pages 30 and 34 are by Dr Neil Clifton (CC BY-SA 2.0) and front cover and page 92 by Alan Armstrong. If for any reason people or organisations have not been accredited as necessary or copyright material has been used without permission/acknowledgement we apologise for any oversight and will make the necessary correction at the first opportunity.

About the Author

Born in Greenwich, London, David Ramzan has always had a keen interest in the history of his home town and studied local history as an undergraduate at the University of Oxford. A self-employed graphic designer, illustrator and artist, he also worked in special educational needs and has written several historical publications and articles for magazines and periodicals, subjects that include local history, maritime and football history, piracy, smuggling and Victorian crime.